GROW

Knowledge for Kids

Alia Wash

Mission: To Proclaim Transformation and Truth

Publisher: Transformed Publishing, Cocoa, FL

Website: www.transformedpublishing.com

Email: transformedpublishing@gmail.com

ISBN: 978-1-953241-44-3

Acknowledgement

Most importantly, I thank God for His healing power

in my body, spirit, and soul.

Dedication

I dedicate this work first to my children, my grandchildren, and their children to come. I pray this book inspires growth in every area of your lives and the lives of the people who will share this book with their children and grandchildren.

Nevaeh, Lonnie, Amir, Ranyila,

Tahiry, Tania, & Kamahl:

Grandma Lia loves you.

I pray this book promotes growth . . .

Special Dedication

To my grandmother,

Carrie Whisonant.

Amir

Time to Go!

Amir was so excited to go to Grandma's house. He knew when he arrived, there were so many exciting things he was going to be doing. He started to daydream about the car ride and predict how long it would take to get there.

Going to Grandma's was the highlight of Amir's year. He knew they would sit around the fireplace, play games, and make his favorite foods. Amir was ecstatic to learn new things about Grandma's animals.

As soon as the last bell of the day rang, Amir quietly pushed in his chair, gathered his jacket and bookbag, and went to stand in line. As the other children lined up behind him, Amir reminded his teacher Ms. Cherry, "As soon as my dad picks me up, we are going to Grandma Lia's house in Georgia."

Ms. Cherry gave Amir a special smile and directed the class to exit the classroom. They reached the door to the parking lot. Ms. Cherry gave all the children a tight hug and walked each child to their parent's car and told them to have a good Spring Break.

When Ms. Cherry got to Amir's dad's car, she opened the door and Amir quickly jumped in and said, "See you in two weeks Ms. Cherry. Off to Grandma's house I go!"

Ms. Cherry looked over at Amir's dad and smiled, "He's really excited, huh?

Amir's dad smiled and said, "Yes, he's been waiting for this week to be over so we can get to Grandma's. He packed his bag two days ago."

Amir cleared his throat and yelled with enthusiasm, "And I'm ready!" Saying goodbye to Ms. Cherry, Amir reached for the door. Amir's dad and Ms. Cherry let out a laugh. They could see Amir was *past ready*. As they pulled off, Ms. Cherry yelled, "Bye, Amir! Enjoy your Spring Break!" as she watched the car leave the parking lot.

To Grandma's House We Go!

Right after they pulled out of the school's driveway, Amir began to sing. "Daddy, do you have my suitcase?"

"Yes, Amir," Amir's dad responded.

Amir continued, "Daddy, do have my boots?"

"Yes, Amir," his dad answered once again.

"Daddy, do you have my cowboy hat?"

"Yes, Amir!" Amir's father replied with a joyful melody, "I got everything, Amir. You're all set lil' man." Amusing himself, Amir's father continued to sing, "Now, let's stop and get you some snacks. We gotta a little bit of a ride ahead of us."

Amir was so excited! He asked his father, "Do we have to stop this time, Daddy?"

"Yes, Amir," his father replied. "This would be a good time to go to the bathroom, too. I know you're excited son. We'll be there soon enough. I brought your favorite books and your blanket so you can relax, and read while I focus on the road."

With a sigh, Amir settled down. They went into the store, bought some snacks, used the restroom, and got back into the car to travel the rest of the way to Grandma's house. As they headed down the road, they sang, "To Grandma's house we go!"

Grandma's House

Two hours later, Amir's dad rubbed Amir on the head to wake him, and announced, "We're here lil' man. We're at Grandma Lia's house. Wake up."

As soon as Amir heard his father, he jumped up out of his sleep, "Grandma! Grandma!" he exclaimed. "I made it!" Amir shouted.

"Yes, we made it, son," his father responded.

They pulled down the long driveway that led them back off the road to a lil' brown house surrounded by trees guiding them to their destination. Amir's eyes lit up!

2

Everything was just how he remembered. The big tall trees, taller than the house. Beautiful different colored flowers that flowed like a stream surrounding the base of the house, like a beautiful rainbow. A tall wire fence went around the back of the house and kept all the animals in Grandma Lia's backyard. It seemed as if they drove across a large football field after they turned onto the property. Finally, coming to a stop in front of the brown house, they made it to Grandma Lia's porch. That's how far the house is from the road.

Amir jumped out of the car with excitement and ran towards the porch where beautiful flowers in enormous pots covered parts of the deck. Flowering plants hung in pots, extending long vines, that went around the porch like a picture frame. There were also two large rocking chairs on the porch and a stack of fire-wood in-between them for Grandma Lia's fireplace. She always keeps wood there. Tiny hummingbirds were flying around, collecting sap from the little hummingbird feeders that were placed on short poles in the ground along the perimeter of the porch, so the birds could easily access them. The beauty of Grandma Lia's house always mesmerized Amir - the flowers, the numerous birds flying around and chirping, and the relaxing sounds of nature as you walk up to the door.

As soon as the car door slammed, the front door of the house opened. There in the doorway was Grandma Lia. Amir crashed into her arms, as she held them open, and called out his name, as he called her's. They tightly held their tender embrace.

Grandma Lia was tall and slim with a caring smile and eyes of wisdom. She wore her flowered overalls, with her matching sun hat, and a pair of worn boots. The same boots she wears, day after day, to work outside in the garden or with the animals. "I've been waiting for you guys," she said proudly, as her and Amir released their long hug. Amir's dad reached the door where they stood and he embraced his mom as well. She welcomely directed them to, "Come on in."

Let's Get Comfortable!

Not only was Grandma's house cool on the outside, but it was even cooler and cozy on the inside. She had more flowers hanging everywhere, oversized furniture, throw blankets all over the couches, a little fireplace to keep the house warm, and family pictures all over the walls, with the smell of cinnamon and sweet spices in the air.

Amir's dad secretly loved coming to visit as well, instantly taking him back to his childhood and all the good times he remembered growing up there.

"How was your drive?" Grandma Lia asked them as they both came in and set their bags down.

"I thought we were never going get here, Grandma," Amir explained.

"Yes, he sang himself to sleep," Amir's dad blurted out. "Singing, 'To Grandma Lia's house we go!'" They all broke out into laughter.

"I'm so glad you guys are here!" Grandma Lia explained, "I've been planning some exciting things for our Spring Break this year, Amir. We are going to grow so much together. I can't wait for you and me to get started, after you guys get some rest. Go get cleaned up for dinner," she instructed them.

4

Grandma Lia loved to cook. She always made her son and Amir's favorite, 15 Bean Soup, when she knew they were coming, loaded up with smoked turkey wings, and crackers on the side

After dinner, they all settled down with a hot cup of chamomile tea as they snuggled on the couch, in front of the fireplace, where they all fell asleep.

First Things First

Waking up at Grandma Lia's was the best! Amir always got up when Grandma got up in the morning. He knew Grandma woke up every morning with the roosters, literally. You can hear the sounds of Grandma's three roosters crowing in her backyard, *cock-a-doodle-doo* every morning.

"Time to start our day," Grandma whispered, as she led Amir quietly to the bathroom to wash his face and brush his teeth. Grandma uses natural soap and toothpaste she makes herself. "Let's stay quiet," she whispered in Amir's ear, "we don't want to wake your dad up. You know he's leaving when he gets up. He has a long drive ahead," Grandma Lia explained to Amir.

"Yes, ma'am. I know," Amir whispered back.

Grandma set up Amir's new toothbrush with black toothpaste. Amir's eyes widen. "Grandma," Amir whispered again, "why is the toothpaste black?"

"It's activated charcoal which is healthier for your little teeth and gums. Grandma makes this toothpaste because it doesn't have all those extra chemicals in it that can affect your little body in a negative way. It's more natural for our bodies and our mouth health."

Amir looked at the toothbrush in amazement. With a nervous look on his face, he began to brush his teeth. His eyes lit up as the black charcoal ran down his mouth. "It's not so bad after all. It makes my mouth feel minty," Amir explained.

"Yes, I know and it's better for my lil' man as well," Grandma replied as she washed Amir's face off.

When they were all done, Grandma led Amir to the kitchen. They sat down at the table, held hands, and closed their eyes together. At Grandma Lia's house, they always thank God for the day He has made.

After prayer, Grandma arose from her seat and said to Amir, "Now let's have our morning tea before we go collect our eggs from the chickens for breakfast."

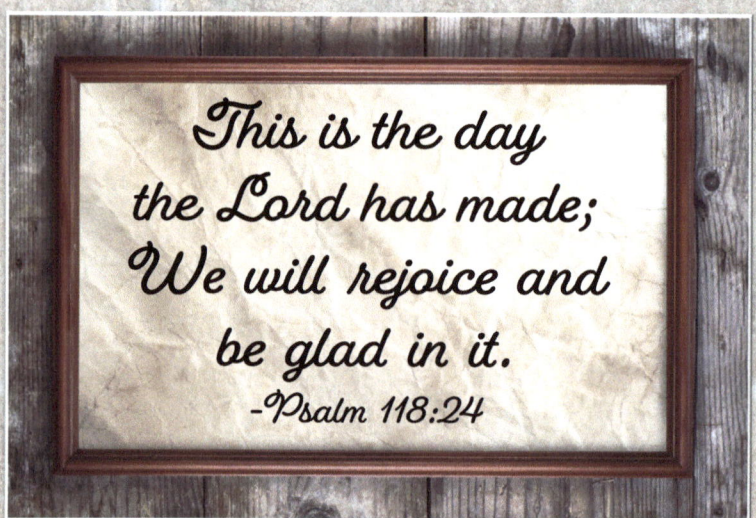

This is the day the Lord has made; We will rejoice and be glad in it.
-Psalm 118:24

Here Chick-Chicky

Once Amir finished his tea, he went to put on his trousers and boots. He proceeded to collect his hat and headed towards the back door where Grandma was waiting. She wanted him to be the first person to open the door. She loved watching his eyes light up to the beauty that was behind it.

As Amir opened the door, the sunshine hit his face as it arose to break dawn. The sun was coming up right before his eyes. Straight ahead of him was a beautiful garden. The tall arch was a decorative passageway. The brilliance of plentiful flowers grew over the arch, and led to a fenced off area with a door to the garden. Grandma called it her outer garden. She fenced it in so the chickens couldn't enter and eat all of her food.

As Amir ran down the porch steps, there was another fenced in garden with all the plants Grandma grew for medicine. Amir descended to the bottom of the steps. He stopped to take all the memorizing wonder of the flowers and plants in. He loved the fact that every time he came to visit Grandma, he got to pick tomatoes, cherries, and muscadines directly from the garden.

Amir looked back at his grandmother's glowing face and said, "Let's get out the baskets to collect the eggs." Amir excitedly grabbed the scooper to feed the chickens.

Grandma Lia giggled, "So, you remember where everything is?"

"Yes, Grandma," Amir responded nonchalantly as he took off running to the shed where all the food and supplies for the animals were kept. There where four coops and two large enclosures for rabbits lined up in a row across from each other on the outskirts of the garden in Grandma's backyard.

"So how many eggs do you think we'll get this morning, Grandma Lia?" Amir asked inquisitively.

"We'll have to count them out and see lil' man," Grandma responded. "I've gotten some more hens since last spring."

Amir's eyes widen, as he ran over to the first pen to throw the feed and collect the eggs. "There are ten hens in there!" Amir cried out, "Here, chick, chick, chicky!" as he threw the food. All the hens ran over in a huddle and began to eat.

"Now, take your time lil' man and go collect our eggs."

"Yes, ma'am," Amir respectfully replied as he ran to check the boxes for eggs. "1, 2, 3, 4, 5, 6, 7, 8, 9, 10," Amir counted out.

"Wooh!" Grandma said in excitement. "Ten chickens in this coop and we got ten eggs! Isn't that amazing, Amir?"

"Yes, Grandma. You know my mommy and daddy hardly buy eggs anymore. My daddy says they're too expensive," Amir explained.

"Yes, I heard that on the news, too. Prices have gone up for eggs and other foods. But don't you worry, lil' one. As long as Grandma got eggs, you got eggs too!"

They collected thirty more eggs from the other two coops, Amir stopped in his tracks and with excitement hollered, "Forty eggs! How will we eat them all, Grandma?"

Grandma smiled, "We're not going to eat them all in one day. We'll have these for breakfast every day while you're here, and the leftovers I'm going to sell to buy more food for the animals."

"Okay, Grandma," responded Amir.

Together, Grandma and Amir hurried along feeding all the animals. Grandma Lia went over to her side garden and collected some oregano and rosemary for their breakfast. She turned on the sprinklers to water the garden and they headed inside to start breakfast.

What's Nutrition?

Amir entered the room in the house that is Grandma Lia's office. Amir studied the room. In all his little years coming to Grandma Lia's house, he never really paid attention to the details of Grandma's office. He just knew Grandma would go in there to get jars off the shelves and take the plants out of them to put in their tea. She would also bag some of the contents of the jars, put them in boxes, and send them to her friends. When Amir would ask her what she was doing, that's what she told him. As Amir looked around the room this time, he noticed he could pronounce the words on the jars. "Grandma! Grandma!" he yelled to Grandma who busy preparing breakfast for them in the kitchen.

"Yes, Amir?"

"I see rosemary in your jars and thyme and moringa! Your herb garden is in these jars Grandma," Amir explained excitingly.

Grandma Lia peeked her head in the room, "Yes, those are all plants from the garden. God created all of those plants to help us maintain a healthy body."

"Grandma," Amir called to her again in amazement, "so why do the plants look dead?"

Grandma Lia started laughing, "They still give life. I put them in the jars after they dry out so I can save all of the nutrients in the plants."

"Nutrients?" Amir turned and asked with a puzzled look on his face.

"Yes," Grandma Lia explained, "it's a substance that provides nourishment essential for growth and the maintenance of life.

"Wow, so why don't we just eat plants, Grandma, if the plants keep us strong and healthy?" Amir inquired.

"Wow, Amir!" Grandma responded, "Come," as she directed him to the bathroom, "Let's go wash those hands of yours. Breakfast is almost ready. After you finish cleaning your hands, come to the kitchen table and I'll tell you all about it."

Grandma was secretly *so* excited inside. She realized Amir was now in a place of understanding. She could pour into him everything he needed to know about why she grows food, herbs, and raises animals.

"Good morning, Daddy," Grandma heard Amir say, as she saw her son emerge through the kitchen door.

"Good morning, Son."

"Good morning, Ma. How's everyone doing this morning?" he asked.

"Good, Son. We are all doing good. God woke us up this morning, so we're doing good." Grandma Lia responded.

"What did I miss?" Amir's dad asked his mother.

"Well, we washed up, said our prayers, tended to the animals, collected our eggs, and now it's time to eat. I'm going to explain to Amir why I have plants in jars," she said giggling, "and why we don't just eat plants." Grandma ended with a lil' sly grin.

"Sounds like you guys did a lot this morning. . . all while I was sleeping," Amir's dad said with a sound of amusement. "I missed out on all of the fun."

Grandma Lia continued to set the table. Amir joined them back in the kitchen and plopped down into his seat. "Grandma, breakfast looks so good!" Amir complemented.

"Yes, Mom this looks great," Amir's father said as well.

"Thank you. Just a lil' something I put together for my boys." On the plate were banana pancakes, scrambled eggs, and turkey sausage. As she sat down to join them, Grandma Lia reached for their hands and said, "Let us pray." They bowed their heads and began to pray.

As soon as they were finished praying, Amir yelled out, "Now, let's eat. I'm starving!" They all laughed and started to eat.

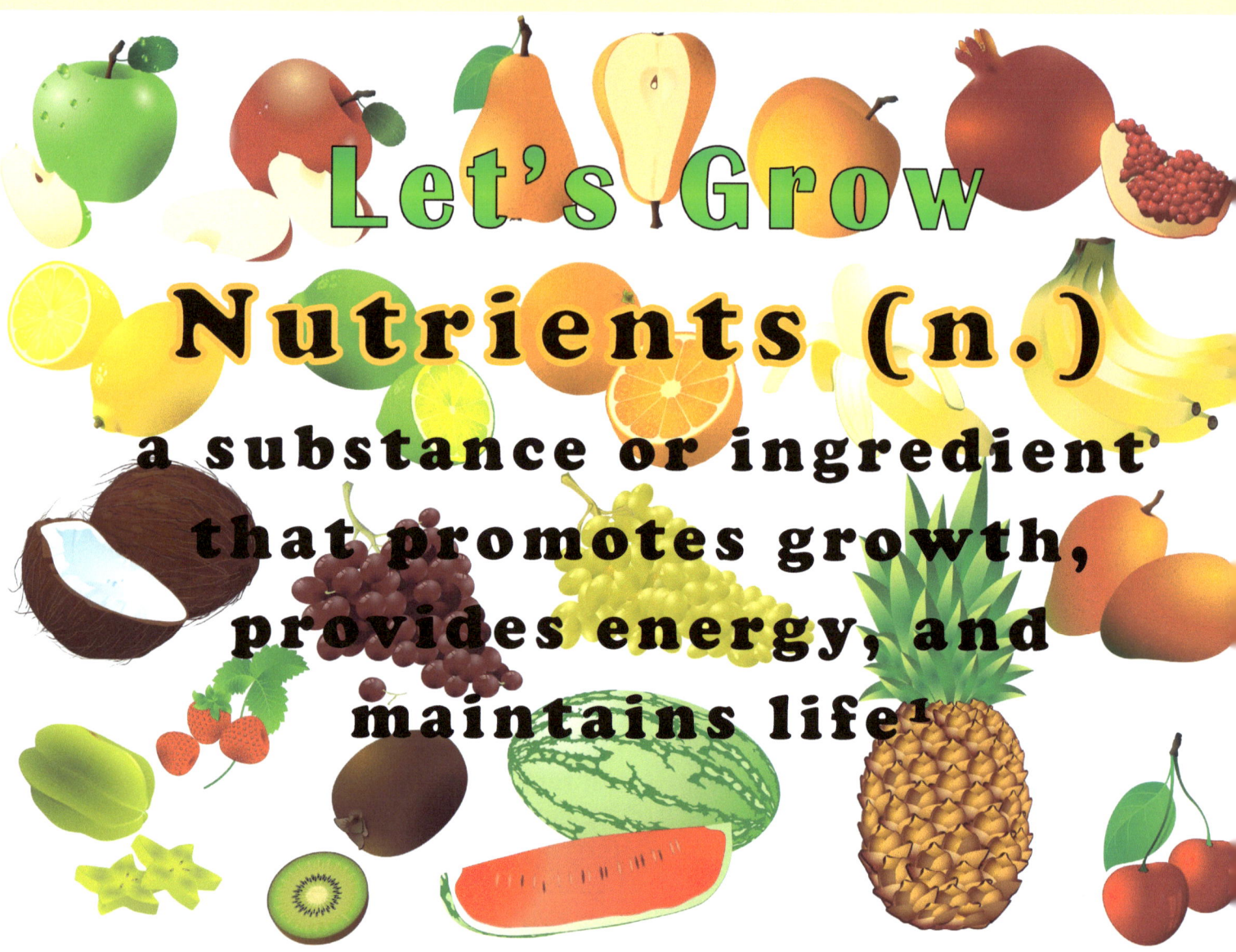

Let's Grow

Nutrients (n.)

a substance or ingredient that promotes growth, provides energy, and maintains life[1]

As they ate their delicious breakfast, Grandma explained to Amir how plants are packed with different nutrients and each individual plant has a different type of nutrient. To answer Amir's previous question, Grandma Lia shared the reason why she doesn't *just* eat plants, "When I was a little girl, meat was introduced to me by my parents and by my school as an important protein. Now that I'm an adult, who has done a lot of research, I know there are other ways to get the protein our bodies need. I don't really eat a lot of meat anymore, only every now and then. We are living beings and we need to eat more living foods to energize our bodies."

Amir did his best to take in all the important information Grandma Lia was telling him. Finally, filled to the brim with Grandma Lia's insight, Amir let out a big, "Huh?" Confused, Amir asked intently, "What are you talking about, Grandma? First, what is protein?"

[1] "Nutrient Definition & Meaning." *Merriam-Webster*, Merriam-Webster, https://www.merriam-webster.com/dictionary/nutrient. Retrieved 3/25/23

"Well, lil' one," Grandma explained, "proteins are basic structures. They are a chain of amino acids. We need protein in our diets to help our bodies repair cells and make new ones. Protein is important for growth and development in children and teens. Ya know, lil' guys like you."

Amir's eyes widen, "Wow, sounds pretty deep," he said as he turned to his dad.

"Yup, Son. Pretty deep. Your grandma has always taught us about our food and *how* food plays a significant role in our health. That's why Daddy rarely ever has a cold," Amir's dad explained, "because I eat a lot of fruits and vegetables, drink lots of water, and eat very little meat just like my mom."

"Wow," Amir replied. I can't wait to tell Ms. Cherry, my teacher, about this. Maybe she'll let me tell my class. Maybe if my friends knew about the Nu – Tree – Ends . . .," Amir stuttered out.

Grandma Lia laughed, "Nutrients lil' man. Nutrients. You'll get it. I think that would be great," Grandma encouraged. "It's important that we educate one another on how to stay healthy."

"Yes, Grandma because a lot of my friends have been out of school this year because they've been sick with a virus," Amir explained.

"I'm so sorry to hear that," Amir's grandma said to him as she grabbed his shoulder. "It really makes me sad to see my friends sick, too."

"Me too, Grandma," Amir replied with his head hung down.

"Don't be sad, lil' man. You're going to learn some new tools this Spring Break that you can go back and share with your friends. You can teach them how they can put more nutrients in their bodies by eating more fruits and vegetables and show them how to grow food."

Amir's eyes lit up. He lifted his head, "Grow food? I'm going to grow food this Spring Break?" Amir inquired excitingly.

"Yes, lil' man," Grandma said as she collected the plates and walked over to the sink. "Let's get everything cleaned up and I'm going to show you how to plant some seedlings now that we will harvest in the summer to keep our bodies healthy and strong. You will be able to tell your friends all about it and maybe even inspire them."

"Yay!" Amir shouted. "Let's get everything cleaned up so we can start our day."

After breakfast, when everything was cleaned up, Amir's dad got his bag and gave Amir a big hug as they walked to the door. Amir's dad looked down at him. "Well, lil' man, enjoy your Spring Break with Grandma. I know you're going to learn some amazing things. It sounds like you're going to be telling the kids about *Nu – Tree – Ends*," as he busted out with laughter.

Yes," Amir said intrigued, "I can't wait to learn so many new things. I love being at Grandma Lia's house!" Amir announced.

"I love you being here with me, too." Grandma Lia responded to Amir as she rubbed his head, then gave her son a hug goodbye, "See you in two weeks, Son." Grandma Lia called out to her son as he stepped off the porch.

"See you in two weeks, Mom. See you in two weeks, Amir. Have fun here with Grandma," Dad said to his mom and Amir.

"I will, Dad. I know I will." Amir said with confidence to his father as he got into the car and backed up out of the long driveway. They stood on the porch and watched Amir's father pull out of the yard.

"Well, are you ready?" Grandma turned to Amir and asked.

"Yes, Grandma. Let's get started planting. I'm so excited!"

"Alright, lil' one. Let's get in the house."

One by One

Amir rushed to the kitchen to find the table already set up. Grandma had removed all of the plates and dishes and laid a blanket over the table. She had all these black cartons on the table, a bag of dirt next to the table, and a box full of seed packages. Amir was surprised because he never saw her put things that are usually outside on the kitchen table. Grandma prepared everything while Amir was helping his dad get ready to leave. "Okay, I'm here!" Amir hollered with anticipation.

Glad to see Amir's excitement, Grandma took the lead, "Let's get started. First, you're going to put your little gloves on, so your hands don't get dirty. Then we are going to fill the black cartons with dirt, pack it down, and take a little pencil to put a hole in the dirt. Finally, we're going to take our seeds and place them in the holes. Each one of these seeds are going to grow into plants."

Amir's eyes lit up as he hurriedly put his gloves on, grabbed his cup, and scooped out some dirt. "This is so fun, Grandma! I never knew I could plant my own food!"

"Yes, Amir. How did you think all of the plants got into the garden?"

"I didn't know, Grandma. I just thought they grew there."

"No, lil' one," Grandma corrected him, "everything starts from a seed. I put the seed in the dirt, water it, and nurture it so it can grow. What we are doing now, is going to provide us with food for the winter. This is where all the food comes from that I send to your house in the box for the winter. All of these seeds grow into beautiful plants that have individual nutrients in them to keep us healthy and strong."

Amir looked over all the seed packages on the table. There were tomatoes, rosemary, oregano, peppers, squash, cucumbers, and more. They planted all their seeds and talked about how long it was going to take for the plants to grow.

When Amir comes back in the summer, he will harvest them with his grandmother as planned. Just the thought of coming back to Grandma's house excited Amir. They spent all morning placing seeds in the dirt. After they finished, they gathered all of their black cartons, with the seeds in them, to take outside to a rack Grandma had in the sun. They placed each black carton on top, then Grandma got the hose and let Amir spray each one with water until they were saturated.

After Amir finished watering, he turned to his grandma and asked, "Now what, Grandma?"

"Now, we'll watch them grow!" Grandma responded.

"I can't wait!" Amir exclaimed with a twinkle in his eye.

"It won't take long, Amir. Just a couple of weeks. But while you're here, we will begin to see them sprout through the dirt. We're going to keep an eye on them."

Grow

Each day was exciting. Amir and Grandma Lia had so much fun together. They spent quiet time playing games, drinking tea, sitting by the fire, and taking care of the animals.

Every morning after Amir and Grandma fed the chickens and watered the animals, they also watered their seedlings. Every day, Amir would turn to his grandmother and say, "Nothing yet!"

Grandma Lia's reply was always the same, "Amir, have a little patience lil' man. Have a little patience.

They repeated the same routine. They got up early every morning, got ready to go outside, said their prayers, drank their tea, fed the animals, collected the eggs, and ate their breakfast. Grandma taught Amir something new each day about the herbs she had in her office.

During the last week of Spring Break, as they progressed through their morning routine, Amir went over to water the seedlings. Astonished, Amir noticed they were sprouting up out of the dirt. He called for his grandmother who was tending to the rabbits, "Grandma! Grandma! They're growing! They're growing!"

Grandma rushed to his side, inspected the plants, and rubbed Amir's head, "Yes, lil' one. They're growing. See, I told you. Just a little patience and they'll grow just like you lil' man. I've been patiently waiting for you to grow so I can teach you all I know about how to grow food and take care of the animals so you can sustain yourself and your family as you become older. This Spring Break has been such a special time with you, Amir. I have something special I want to show you. Come with me over to the animals." Grandma said to Amir as she led him over to the rabbit pen. Grandma pointed to one of the rabbit cages, "Look inside."

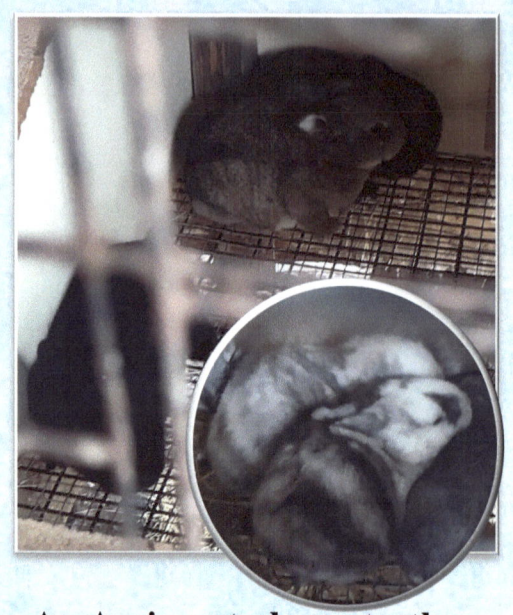

Amir walked over to the cage. He saw a fluffy brown rabbit in the cage and a box in it. "Why does the cage have a box in it, Grandma? Is that her bed?"

"No, Amir. That's her brooder box."

"Her beauty box?" Amir sillily proposed.

"No, Amir," Grandma said laughing. "Her brooder box. That's the box I put in there when she's pregnant and going to have babies.

"She's pregnant?" Amir inquired with excitement.

"She was pregnant. Amir, look a little closer."

As Amir got closer to the cage, he peered into the box. He was amazed when he saw fur moving around in the box, "Oh my goodness, Grandma! There are babies in there!"

"Yes, Amir. We got babies!" Grandma Lia shouted.

"Wow, I've been taking care of them all this week and I didn't even know she was pregnant," Amir realized.

"Yes, that's the surprise I had for you. She's been pregnant for thirty-two days and I knew the babies would be born while you were here visiting. I wanted to keep it a surprise for you." Amir's gigantic smile was duplicated on Grandma Lia's face as they observed the babies in awe.

"This is a grand surprise, Grandma," Amir emphasized. "May I take one of them home with me, Grandma Lia. . . please."

"No, not yet," Grandma Lia explained to Amir, "but as soon as they're old enough, I'm going to bring one to you."

"I can't wait! So, what do we feed them?" Amir asked eager to help.

"We don't feed them. Their mom feeds them, and we feed her. All she's going to give them is her milk, and that is all they need," Grandma Lia explained, "So, just like we need nutrients to be healthy and strong, so do the animals. We're going to give the mama rabbit some kale, oatmeal, and Timothy Hay. That will provide her with the nutrients she needs to feed her babies so they can grow big and strong. Then you will be able to keep one at your house."

Amir shouted full of anticipation, "I can't wait! Hurry up and grow little baby rabbits - hurry up and grow!"

"What did I tell you, lil' man? Have a little patience. Have a little patience. Everything needs time to grow," Grandma emphasized.

"Yes, ma'am Grandma Lia. Yes ma'am.

Time to Go

Spring Break has come to an end. It's the last day Amir will be at Grandma Lia's house. Amir's dad called to say he was on his way. Grandma Lia and Amir got up that morning and went through their routine. This time, when they checked on the seedlings, they had sprouted out from the dirt, and stood strong. They were ready to go into the greenhouse.

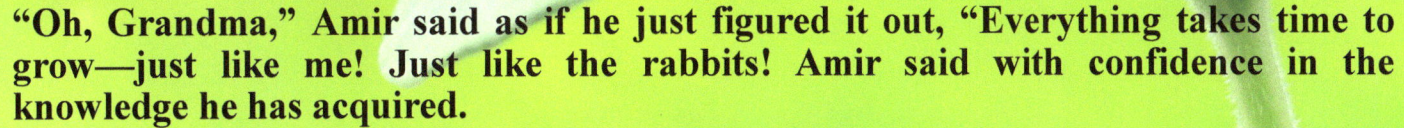

"I'm so excited! The seeds have grown over the last two weeks just from us putting them in the dirt," Amir declared.

Together, they gathered all the seedlings and placed them in the greenhouse. "Why are we putting them in the greenhouse?" Amir asked his Grandma Lia.

"So, they can continue to grow big and strong, away from the elements. Eventually, I'll place them in the garden."

"Oh, Grandma," Amir said as if he just figured it out, "Everything takes time to grow—just like me! Just like the rabbits! Amir said with confidence in the knowledge he has acquired.

"Yes, Amir. Everything takes time to grow, and you have grown so much over this Spring Break. I enjoy spending time with you lil' man and teaching you everything I know. One day you're going to grow up like me and you're going to have a family. You're going to teach your family how to grow their own food and how to take care of their bodies naturally with what they grow," Grandma Lia said to Amir with sadness in her eyes.

Amir looked up at his grandmother to address the sadness in her eyes, "Grandma, why are you sad?"

"Well, lil' one, I hate to see our time come to an end. I know you'll be back in the summer, but I hate to see you go," Grandma explained to him.

Amir grabbed his grandma's hand, as he looked straight into her eyes, "Patience, Grandma Lia," Amir said with care and concern in his own eyes. "Only a couple of months to go and I'll be back for the summer so we can harvest everything we have grown. Plus, I got to get back to school to tell everybody everything I have learned at your house this Spring Break," Amir assured his beloved grandmother as he gave her a huge tight hug.

"Yes, lil' man. You got to let them all know what you learned so they can have an opportunity to grow too by learning something new," Grandma Lia encouraged Amir. "Now, let's go inside and get your stuff together. Your dad will be here soon."

Amir went to his room in Grandma Lia's house. He packed up all his stuff - his hat, his blanket, his clothes, and his boots then he set his suitcase by the door.

Amir and Grandma Lia sat by the fire and drank a cup of tea together for the last time this Spring Break. They went over all they had done while Amir was there. Grandma Lia reminded him she was going to be coming to visit and bring his baby rabbit to him. They snuggled and giggled until there was a knock at the door. It was Amir's dad coming to pick him up.

Amir jumped up to open the door for his dad. He immediately gave his dad a big hug and greeted him, followed by Grandma Lia welcoming her son. Amir was so excited! He told his dad all the things they had done over Spring Break. How they grew food, cared for baby rabbits, and learned about herbs and what is good for his body. They sat and talked awhile as Grandma Lia made dinner.

After eating dinner, they loaded the car as Grandma Lia stood at the door of the little brown house. She blew kisses to them, as they pulled out of the long driveway.

Grandma Lia closed the door and sat in her rocking chair by the fire, drank her tea, and thought to herself, *My how my lil' man has grown.*

Back to School

The first day back to school from Spring Break, Amir's class had Show and Tell. Amir brought a pot of seedlings he planted at Grandma Lia's house over Spring Break. When it was Amir's turn to do Show and Tell, he explained to all of his classmates how he planted *this* cucumber plant with his grandmother. He told them what she taught him about all the nutrients in plants and how good they are for our bodies. How vegetables and fruits give our bodies the fuel we need to be healthy and prevent sickness. Amir's classmates were engaged and had so many questions Amir was able to answer. Amir felt so good as he shared such important information. He even announced that he was expecting a baby rabbit once it got bigger and Grandma Lia was going to personally deliver it when she comes to visit him.

After Show and Tell, Amir placed his plant on the windowsill in his classroom. Each day Amir and his classmates watched the cucumber plant grow. They were all so excited. Soon little flowers popped up on the plant and eventually produced cucumbers. With some more patience, the cucumbers grew to maturity. Ms. Cherry harvested the cucumbers by cutting them off the plant. Then she sliced one up for everyone to have a piece. They ate it in amazement and agreed, *Oh, how good this cucumber tasted!* Many of the students said they were going to tell their parents, "We should grow some food. It tastes so good and it's so easy to do."

Ms. Cherry walked over to Amir, rubbed his head, and whispered, "Amir, I'm so proud of you for teaching the class everything you learned at Grandma Lia's house. Now, we can all learn how to grow our own food. I'm so proud of you!" Amir looked up at Ms. Cherry with widen eyes as she continued, "Amir, my how you've grown."

About the Author

Alia Wash is a wife, mother, grandmother, peer specialist, author, recovery advocate, and a self-taught herbalist. She was born in Newark, NJ in 1977 and was raised by her grandmother, from the age of three years old, in Cocoa, FL. She remembers, as a young child, spending time with her grandmother tending to her garden. At the time, she didn't realize how those years with Grandma created such a great impact in her life. Alia fondly remembers those foundational days when they lived a more self-sufficient life. Most of their diet was vegetables from the garden and wildlife her grandmother captured.

Later on in years, Alia's grandmother was diagnosed with breast cancer and given six months to live. She chose to use all natural remedies, without the help of doctors, and lived an additional fifteen years, far exceeding the given prognosis. This remarkable testimony was embedded in Alia's mind for most of her childhood into her adult life.

Alia was plagued with different types of diseases. After fifteen years of suffering from lupus, fibromyalgia, and diabetes, Alia put into practice the experiences she had with her grandmother throughout her childhood. She remembered the things her grandmother had taught her and how she lived longer than the doctor had predicted. Embarking upon her own journey, Alia was able to heal herself from lupus, fibromyalgia, and diabetes. She no longer takes medication for these illnesses and has begun to help others heal themselves through natural medicine.

Alia continues her grandmother's legacy of hard work, research, and faith as she diligently pursues her own personal journey of health and agriculture to become more self-sufficient and heal herself naturally. It is her mission to reach back and help others learn how to do the same. But most importantly, to help the children of our future.

Learn more at: https://heavenscurenaturalfoods.com/

Or Email - Homesteading with the Washes:
aliawash6@gmail.com

Learn more about Alia's story in her first book titled, *Why Not Recover*, 2018.

Why Not Recover, is a tool designed to assist those who desire to relieve themselves from the bondage of addiction and the cycle of relapse.
In this book you will receive the keys to obtain and maintain your freedom.

Available on Amazon.com.